GROW
Cycads

A GUIDE TO THE SPECIES, PROPAGATION AND CULTIVATION OF SOUTH AFRICAN CYCADS

Text and photographs by
John Donaldson and John Winter

Right: Mature *Encephalartos* species at Kirstenbosch

Below: The striking male cones of *Encephalartos woodii*, a cycad that is extinct in the wild

CONTENTS

Introduction 5

Propagation from seed 9
- Collection of pollen
- Pollination of the female cone
- Harvesting and storage of seed
- Sowing and germination
- Seed sowing mixes

Vegetative propagation 15

Cultivation 17
- Soil
- Planting
- Watering
- Transplanting
- Feeding

Container plants 23
- Growing medium
- Finding the right position
- Watering

Pests and diseases 25

Landscaping with cycads 29

Cycads and the law 33
- Do you need a permit?
- How do you know that your cycad is cultivated and not collected from the wild?

Further reading 35

Index 36

Opposite: The feathery crown of *Encephalartos ghellinckii*

Below: *Encephalartos aemulans* in its rugged natural environment

INTRODUCTION

If one takes the world's flowering plants with 250 000 species, the cycads are a small group comprising 11 genera and approximately 250 species. The main centres of distribution are Central America (4 genera), Australia (4 genera) and southern Africa (2 genera). The African genus, *Encephalartos*, is the second largest in the cycad family and consists of about 60 species with the main centre of distribution in the eastern parts of South Africa where 37 species occur. The smallest genus, *Stangeria*, is also endemic to South Africa and consists of a single species that resembles a large fern.

African people have had a long-standing interest in indigenous cycads, using them in medicinal and cultural practices and occasionally planting them around the homestead as part of traditional customs or purely for their enjoyment. The rest of the world can trace their exposure to South African cycads initially to Carl Peter Thunberg and Francis Masson who undertook various botanical expeditions across southern Africa and came across *Encephalartos caffer* and *Encephalartos longifolius* in the Eastern Cape in 1772 and 1773. As followers of the Linnaean tradition of identifying and naming plants, they recognized the uniqueness of these cycads and shipped specimens to Europe. Other species followed, and soon cycads

C Giddy

were being dispatched from South Africa to gardens all over the world. So began a fascination with growing these unusual plants that still flourishes in many parts of the world.

Although sometimes regarded as quaint palm-like relics from an ancient era, cycads are far more varied and versatile than their reputation as 'living fossils' suggests. The South African species exhibit a wide range of growth forms and colours and have great horticultural appeal. In the past, enthusiasts and gardeners would simply collect cycads from the wild. Sadly, such collecting led to the collapse of many natural populations and conservation laws were introduced to prohibit the removal of plants from the wild. Fortunately, cycads are relatively easy to grow from seed and it is quite unnecessary to plunder the last remaining natural populations. Despite their reputation for slow growth, reasonably sized cycads are generally available from nurseries and small plants can flourish given the right growth conditions. Creative landscaping can also show off the attractive features of smaller cycads while giving them space and time to mature.

Growing your own cycads involves finding out about a range of subjects dealt with in this book:
■ how to propagate cycads from seeds and suckers?

- what soil mixes and fertilizers are needed for cultivation?
- how to recognise and treat pests and diseases?
- where to plant cycads and how to use them in landscaping?
- what permits are required if you own a cycad?

Opposite: Male and female cones in a cluster of *Stangeria eriopus*, a small cycad with a subterranean stem

Right above: *Encephalartos horridus*

Right: *Encephalartos senticosus* overlooking a canyon in northern Kwazulu-Natal

Right: The ripe female cone of *Encephalartos longifolius*

Below: Male cones of *Encephalartos villosus*

Opposite: The bright fleshy seed coat of *Encephalartos caffer*

PROPAGATION FROM SEED

Cycads can be propagated from seed and vegetatively by removing suckers from the main stem of the mother plant.

Male and female cycad cones are borne on separate plants. To produce fertile seed from cultivated cycads, it is necessary to hand pollinate the female plants. The only way to tell a male plant from a female plant is by comparing their cones: the male cone is usually long and slender with small cone scales while the female cone is egg-shaped with large cone scales.

Collection of pollen

Just before pollen is shed, the male cone starts lengthening. This is easily noticed as the tightly packed scales start separating. Pollen starts being shed once the central axis is fully extended. Remove the cone at this stage by covering it with a plastic bag and cutting it off at the base. Place it on a piece of smooth paper in a draught-free room. After a few days all the pollen will have been shed and it can be used immediately for pollination or stored for later use. Before storing, pass the pollen through a fine sieve to remove all impurities and place in a sterilized jar. The lid should be sealed to prevent moisture reaching the pollen. A sachet of 'silica gel' placed in the jar will absorb any excess

The fine herring-bone pattern of leaves of *Encephalartos lebomboensis*

Opposite: Dry method of hand pollination. Notice the 'cap' of scales to the right of the cone

moisture. Store the pollen jars in a deep freeze. This pollen can be used successfully for a year or two.

Pollination of the female cone
When the female cone is mature, scales on the top half of the cone separate slightly and move about 2 to 3mm away from one another, and it is assumed that the female cone is now receptive. Despite these changes it is often difficult to know when cones of some species are receptive.

If there are male cones in the vicinity, natural pollination may occur. However, to ensure good seed set and to avoid hybridisation, it is better to collect pollen and to artificially pollinate female cones.

There are various methods of artificial pollination. For best results, pollinate the female cone as many times as is possible, regardless of the method, using 1tsp pollen each time.

Method 1 Using a teaspoon, disc or spatula place the pollen in or in front of the openings between the scales of the cone. Use a small spray or syringe to blow the pollen into the cone at all of the visible openings. This process should be repeated daily until the spaces between the scales close.

Alternatively, remove the uppermost scales with a sharp knife and inject the pollen into the spaces next to the cone axis. An infant enema is ideal for this purpose. After pollination, replace the 'cap' of scales on top of the cone.

Method 2 Take 1tsp pollen and mix it with 200ml distilled water. Pour the mixture into

the openings between the scales. Alternatively squirt the solution into the openings using a syringe with a narrow ridged tube attached. Repeat daily until the openings between the scales close.

Method 3 Place a plastic bag over the female cone and tie at the base of the cone. Using a plastic syringe, puncture the bag at the top and blow the pollen into the bag with the syringe. Remove the bag after 15 minutes and repeat the process daily until the openings between the scales close.

Harvesting and storage of seed

It is believed that the flesh covering the seed contains carcinogens and precautions should be taken when harvesting and cleaning cycad seed. Handle seeds in a well ventilated room or outdoors, and wear a face mask and gloves. When mature, the female cone begins to disintegrate and the entire cone should be removed from the plant. Immerse the seeds in water for a few days to soften the flesh. Rinse seed in clean water to ensure that all flesh is removed before treating with a fungicide (eg Benlate or Kaptan). Once dry, store the seed in brown paper bags in a cool dry place.

Sowing and germination

After harvesting, cycad seed should not be sown immediately, but stored for a period of at least six months. From the time of fertilization the ovule develops continuously and even when the seed is harvested, the ovule continues to develop. The seed is ready to germinate after six to nine months, when the embryo is fully developed.

To germinate, cycad seeds require moisture, warmth, humidity, plenty of oxygen and a clean, healthy environment.

Bottom heat (28°C) stimulates germination although cycad seed will germinate at a lower temperature, but the process will be slower. Regular watering helps to maintain relatively high humidity levels and a well aerated propagation medium containing coarse particles provides good aeration and drainage. Fungi and bacteria can destroy cycad seed, so as a precaution before sowing, treat the seed with a fungicide solution (eg Kaptan) and discard all floating seeds. These floating seeds indicate that the embryo has not developed and they are infertile.

The conventional method of sowing is to place the seeds in rows lying horizontally next to one another on the propagation

Close-up of a male cone at pollen-shed. Note the separated cone scales with pollen sacs on the lower surface

Opposite: The blue-grey leaves and pale green cones of *Encephalartos middelburgensis*

Right: Close-up of the large cone scales of a female *Encephalartos*

bed or in containers. The seed is pressed down level with the surface of the propagation medium. It is important to label the seed with the name and the date of sowing.

The propagation bed should be 200mm deep to allow sufficient depth for root development. Seed should be sown six to nine months after harvesting. Germination can take place within a month or longer with the root appearing first. Once the first leaf has developed, the young seedling can be planted into a *2l* bag and placed in light shade.

Seed sowing mixes

The most suitable medium for germinating cycad seed is one that is coarse and that will allow drainage yet be able to retain adequate moisture for germination. The sand should be washed and, as sand has a tendency to dry out, milled pine bark or peat added in the ratio of two parts sand with either one part milled pine bark or peat.

Above: Seedling of
Encephalartos villosus

VEGETATIVE PROPAGATION

The only method of vegetative propagation of cycads is by division. This entails the removal of the suckers or 'offsets' that usually occur at the base of the mother plant. Use a clean sharp spade, or knife to remove the suckers.

The best time to remove suckers is in early spring. They should be larger than 100mm in diameter and should be allowed to develop to as large a size as possible before being removed. The larger the sucker the greater the possibility of success. Before removing the suckers, prune all their leaves to reduce transpiration (loss of moisture through the leaves) and make them easier to reach. When removing a sucker the point of attachment must be exposed and the sucker severed with a sharp knife or spade causing minimal damage. All cut surfaces should be treated with flowers of sulphur to prevent fungal infection. Store the sucker in a cool dry place for three weeks to allow the wound to seal.

Before planting in a well-drained growing medium, dust the wound with a rooting hormone to stimulate root development. Large suckers can be planted directly into the open ground provided the soil is well drained. Smaller suckers can be planted in containers, placed in light shade and kept moist. When the sucker has produced leaves, regular feeding is recommended once a fortnight with a balanced liquid fertilizer during the summer months.

Right: *Encephalartos altensteinii* at Kirstenbosch

Below: *Encephalartos villosus* (foreground) does best in shady parts of the garden

CULTIVATION

Cycads take well to cultivation provided they are grown in conditions similar to their natural habitat. For example *Encephalartos villosus* occurs in forests with mild winters and will therefore not do well in regions where the winters are severe.

Soil

Cycads tolerate a wide range of soils. However, regardless of the type of soil, the key factor for the successful growing of cycads is that the soil be well drained. Cycads can even be grown successfully in clay soils, provided that adequate drainage is provided. Adding gypsum and organic material will help to break down the clay and improve aeration. Sand is generally deficient in organic material, low in plant nutrients and dries out rapidly. Growing cycads in sandy soil requires the addition of organic material in the form of manure or mature compost applied regularly as a mulch. This will greatly improve the nutrient content and help to retain moisture. Loam soils are usually ideal for growing plants, as they are fertile, well drained and do not dry out rapidly. A pH of between 7 and 8 is ideal.

Planting

Cycads in containers respond well to being planted into the open ground, as they prefer a free root run. Prepare a

Distribution, availability, features and growing conditions for South African cycads

W moderate watering
L low water requirements
S subterranean stem
T tall stems, tree-like
ss short stem with suckers
B blue grey leaves

G green leaves
N narrow leaflets

○ full sun
◐ semi-shade
● deep shade

🌿 generally available at nurseries
🌿🌿 available at specialist nurseries
🌿🌿🌿 rarely available

✻ frost hardy
☆ moderately frost hardy

Encephalartos

Species	Description	Availability	Water	Leaves	Stem	Hardiness	Sun
E. aemulans	on rocky slopes in northern KwaZulu-Natal	🌿🌿🌿	W	G	T		○
E. altensteinii	in forests and grasslands in the Eastern Cape	🌿	W	G	T		○◐
E. aplanatus	in shaded woodland, eastern Swaziland	🌿🌿🌿	W	G	S		●
E. arenarius	on sandy slopes in the Eastern Cape	🌿🌿🌿	L	G	ss		○◐
E. brevifoliolatus	very rare, on grassy slopes, Northern Province	🌿🌿🌿	W	N	T	✻	○
E. caffer	dwarf cycad in grasslands of the Eastern Cape	🌿🌿	W	G	S		○
E. cerinus	rare dwarf cycad, on cliffs and among rocks, northern KwaZulu-Natal	🌿🌿🌿	W	G	S		○
E. cupidus	small cycad, rocky soil, Mpumalanga	🌿🌿🌿	L	B	ss		○
E. cycadifolius	on rocky slopes in the Winterberg mountains	🌿🌿	L	N	ss	✻	○
E. dolomiticus	rare, on dolomite ridges in the Northern Province	🌿🌿🌿	L	B	T	☆	○
E. dyerianus	rare, on rocky slopes in the Northern Province	🌿🌿🌿	L	B	T	☆	○
E. eugene-maraisii	rocky soils in the Waterberg mountains, Northern Province	🌿🌿	L	B	T	☆	○
E. ferox	in sandy soils, northern KwaZulu-Natal	🌿	W	G	ss		○
E. friderici-guilielmi	high altitude grasslands, Eastern Cape and KwaZulu-Natal	🌿	L	N	T	✻	○
E. ghellinckii	in grasslands, up to Drakensberg foothills	🌿🌿	W	N	T	✻	○
E. heenanii	rare, on rocky slopes in Mpumalanga and Swaziland	🌿🌿🌿	W	G	T		○
E. hirsutus	in rocky terrain, Northern Province	🌿🌿🌿	W	B	T		○
E. horridus	in rocky soils, semi-arid parts of the Eastern Cape	🌿🌿	L	B	ss	☆	○
E. humilis	in grasslands, Mpumalanga	🌿🌿	W	N	ss	✻	○
E. inopinus	rare, on rockly slopes in semi-arid areas of Northern Province	🌿🌿🌿	L	B	ss	☆	○

E. laevifolius	rare, on grassy slopes, Mpumalanga, Northern Province, Swaziland	🌿🌿🌿 W	N	T	❋	○	
E. lanatus	on rocky outcrops in Mpumalanga and Gauteng	🌿🌿 W	N	T	❋	○	
E. latifrons	very rare, in thicket and among rocks in Eastern Cape	🌿🌿🌿 W	G	T		○	
E. lebomboensis	on rocky slopes in KwaZulu-Natal and Mpumalanga	🌿 W	G	T		○	
E. lehmannii	in stony soils on the eastern edge of the arid Karoo	🌿🌿 L	B	ss	☆	○	
E. longifolius	on rocky slopes and in sandstone, Eastern Cape	🌿 W	G	T		○	
E. middelburgensis	rare, on slopes and among rocks on the highveld, Mpumalanga	🌿🌿🌿 L	B	T	☆	○	
E. msinganus	rare, on rocky slopes in a dry area of KwaZulu-Natal	🌿🌿🌿 L	G	T	☆	○	
E. natalensis	widespread on rocky slopes in KwaZulu-Natal	🌿 W	G	T	☆	◐	
E. ngoyanus	dwarf cycad, among large boulders in northern KwaZulu-Natal	🌿🌿 W	G	S		◐	
E. nubimontanus	rare, on dry rocky slopes in the Northern Province	🌿🌿🌿 L	B	T		○	
E. paucidentatus	in dense shade and in open areas, Mpumalanga	🌿🌿 W	G	T		◐	
E. princeps	on rocky slopes in semi-arid areas of the Eastern Cape	🌿🌿 L	B	T	☆	○	
E. senticosus	on rocky slopes, northern KwaZulu-Natal and Swaziland	🌿 W	G	T		○	
E. transvenosus	on hillsides in the Northern Province, sunny to dense shade	🌿 W	G	T		◐	
E. trispinosus	in rocky soils, semi-arid areas of the Eastern Cape	🌿🌿 L	B	ss	☆	○	
E. umbeluziensis	hot, dry conditions in low scrub, Swaziland, Mozambique	🌿🌿🌿 L	G	S		◐	
E. villosus	in densely shaded forests, Eastern Cape to Swaziland	🌿 W	G	S		●	
E. woodii	originally from the Ngoye Forest, KwaZulu-Natal, dense shade, now extinct in the wild	🌿🌿🌿 W	G	T		◐	

Stangeria

S. eriopus	two forms, one from grasslands, one from forests, Eastern Cape and KwaZulu-Natal	🌿🌿 W	G	S		◐

generous sized hole of 500 x 500 mm square and deep. Mix in liberal amounts of old manure or mature compost and two cups of bone meal. Before removing the cycad from the container tie the leaves together to facilitate handling. Water the plant well before removing from the container as this will enable the plant to slide out easily. Place the cycad in the hole just below ground level and firm the soil down around the plant. Use the surplus soil to create a basin around the plant to retain water.

Watering

In their natural habitat, South African cycads require a rainfall of 375 mm to 1250 mm per year. It is important that the water requirements of the various species be taken into account (see table).

Cycads are able to withstand relatively long dry spells, although plants that receive water regularly appear healthy and do not show signs of stress (eg yellowing and reduction in the size of leaves) if they are kept in good condition. Mulching around the plants with a 100 mm thick layer of compost or manure retains moisture and reduces weed development.

During the summer months, give sufficient water to thoroughly soak the root area once a week. There is no need to water the plants in winter as most cycads occur naturally in summer-rainfall regions. When operating automatic watering systems in gardens, it is essential to establish two watering regimes: firstly, a general garden routine which would include most of the cycads and secondly, a special routine for cycads that must be treated as succulents and watered accordingly. (See table page 18)

Remember that good drainage is essential!

Transplanting

The term transplanting refers to moving plants growing in the open ground from one position to another. Cycads transplant relatively easily, although there are a number of exceptions: *Encephalartos cycadifolius, E. ghellinckii, E. lanatus* and *E. paucidentatus* do not transplant well as adult plants.

The best time to transplant cycads is in early spring just before their growth-cycle begins. Transplant preferably during the plant's rest period (winter to early spring). Remove all of the leaves as this reduces the loss of moisture from the plant while it is recovering from the transplanting and reduces stress. Dig a circular trench around the cycad (removal of the leaves will make this easier). Sever the roots with a sharp spade while digging the circular trench, but try to retain as much root as possible. Prepare the new site by adding mature compost and old manure to the planting hole. Ensure good drainage. Plant at the same level as it was previously planted and water well, then water sparingly until new leaves appear. This could take one year or more. Very tall plants should be secured with stays until well rooted. If the plant is not planted immediately, the root ball needs to be wrapped in hessian to prevent the roots from drying out. Mature plants can remain out of the soil for long periods (± 6 months) provided they are stored in the shade.

Feeding

It is essential to feed and water cycads regularly in order to maintain growth and good quality plants. Cycads are long lived and have the reputation of being slow growers. Some are, however, faster growing than others and under favourable conditions will grow accordingly well.

A balanced inorganic fertilizer should be applied twice a year during the growing season at the rate of 1 kg to each mature plant (eg 3:1:5 or slow release fertilizer). This will keep the plant in good condition. In addition a mulch of well rotted compost or manure applied once a year around the base of the plant will improve the condition of the soil and provide extra nutrition.

Cycad nursery at the Lowveld National Botanical Garden

Opposite: A garden landscape using *Encephalartos longifolius, E.altensteinii* and agaves

A mature and unusually tall specimen of *Encephalartos humilis*

Right: *Encephalartos ferox* as a container subject

CONTAINER PLANTS

Cycads make suitable container plants that can be used successfully in displays. They are ideal outdoor container plants provided they are positioned where there is adequate space to avoid their sharp spines. The only South African cycad suitable for indoor display is *Stangeria eriopus*, which produces attractive foliage when grown in shady conditions. Terracotta pots look most attractive but are porous and dry out very quickly, unlike concrete containers. A large vigorous species such as *Encephalartos altensteinii* will make an excellent container plant but will have to be planted in the open ground eventually as it will become too large.

Growing medium
A good soil mix with good aeration, moisture retention and nutrition is important for successfully growing cycads in containers. It should contain:
- 6 shovels coarse sand
- 8 shovels milled bark
- 6 shovels friable loam
- 80 g dolomitic lime
- 6 g iron sulphate
- 460 g osmocote 16-18 months

Before potting, the soil mixture must be thoroughly mixed and moist. Fast growing species will develop a root system that will fill the container very quickly. These plants will need to be repotted into larger pots as they develop, otherwise their expanding root system will eventually burst the pot. Slow release fertilizers (eg Osmocote) are ideal for feeding container plants.

Finding the right position
The positioning of cycads is important with respect to exposure to wind, sun and frost. Shade loving species will be able to tolerate early morning sun, but shading will be required during the heat of the day. Exposure to strong wind slows down the growth rate and stunts growth, although hardy species are able to adapt if soil, nutrition and water requirements are adequate. The advantage of growing cycads in pots is that if they fail to thrive in one place, they can be moved to a more suitable position. Pots should be raised from the ground to improve drainage and to prevent the cycad from rooting through into the ground. Frost limits the choice of cycads, but plants in containers can be moved into protected, sheltered positions during frost periods.

Cycads for containers
E. altensteinii	E. arenarius
E. ferox	E. horridus
E. inopinus	E. latifrons
E. lebomboensis	E. lehmannii
E. longifolius	E. natalensis
E. princeps	E. transvenosus
E. trispinosus	E. umbeluziensis

Watering
Although growing cycads in containers has great appeal, it is labour intensive as plants dry out rapidly. To maintain a container grown cycad in a healthy condition, it is essential to water every other day in the summer months. Container plants become fairly root-bound and in hot weather or windy conditions, they need to be watered regularly and never allowed to dry out completely. The growing medium must be kept moist and surplus water must be allowed to drain out to prevent rotting and help rid the soil of excess salts. The build-up of salts is harmful to plants. Hosing down the foliage regularly gets rid of dust and discourages red spider mite and mealy bug.

Left: *Encephalartos ghellinckii* growing wild in the Drakensberg

Opposite above left: Leopard Magpie moth caterpillars (*Zerenopsis leopardina*)

Opposite above right: Cycad weevil, *Antliarhinus peglerae*

Right: Female cycad weevil (*Antliarhinus zamiae*)

PESTS AND DISEASES

South Africa's indigenous cycads are host to numerous insects, more so than cycads anywhere else in the world. In their wild state, these insects are a valuable component of the country's biological diversity and represent a rich history of interaction between plants and insects dating back many millions of years. When cycads are grown in cultivation, some of the local cycad insects can become pests, joining the ever-present garden pests like mealy bugs and aphids. Fortunately, relatively few insects are real problems and it is only necessary to control them under certain circumstances.

Caterpillars The caterpillars of four geometrid moths (loopers) occur on cycads in South Africa. Two of these, the leopard magpie moth (*Zerenopsis leopardina*) and the dimorphic tiger (*Callioratis abraxas*) are common pests on garden plants. The leopard magpie moth is unmistakable – the day-flying adults are deep orange with black wing spots and they lay clusters of bright yellow eggs on newly emerging leaves. The voracious orange and black larvae feed in large groups on the young leaves and may also eat older leaves and even cones once the new leaves are decimated. In contrast, caterpillars of the dimorphic tiger are cryptic, grey brown, and occur singly on older leaves. The first signs of damage by the dimorphic tiger is an oblong hole (5mm long) in the centre of the leaflet with a grey brown margin. As the caterpillars grow, they feed on the leaf margin and the result is leaves with jagged grey brown edges.

To control, catch the moths as they are laying their eggs and remove egg clusters from the leaves. You could also spray the leaves with a contact insecticide (eg Malathion) preferably before the caterpillars reach 5mm in length as after this they cause extensive damage.

Midges One of the most devastating pests is an unnamed gall midge that attacks young leaves. In low numbers, the first sign of damage is sickle shaped leaflets usually with a brown spot on the inner margin. In higher numbers the midges kill the young leaves before they have properly emerged – the blackened remains make the leaves look as though they have been burned. Eggs are laid on the leaves as they emerge from the crown of the plant. The tiny orange larvae (2mm) develop quickly and are found only on very young leaves where they feed on the inner margins of the leaflets. Usually by the time the symptoms are observed, the larvae have already pupated.

Careful control is essential. First remove all damaged leaves and then use a combination of contact insecticide and trans laminar (across the leaf) insecticide. At Kirstenbosch, a combination of Malathion and Dipterex applied to the crown and to emerging leaves gives good results.

Beetles Two snout beetles, *Antliarhinus zamiae* and *A. signatus*, feed only on cycad seeds. They are spectacular insects with long needle-like snouts.

A. zamiae has the longest snout relative to body length for any known beetle. If you are not harvesting seed, do not worry about these beetles. If you are propagating plants from seed, these beetles can be a serious pest. They lay their eggs into the cycad ovule at the time of pollination (March – June), develop inside the ovule and emerge when the seeds fall from the cone. As many as forty larvae develop in each seed and a single cone can produce more than 5000 beetles.

Other beetles found in the cone are not harmful and should be left alone. This is also true of other insects found in the cones such as cockroaches.

Beetles can also damage the cycad stem. *Phacecorynes variegatus* has a velvety brown sheen with small black marks similar to the Ace of Spades. The larvae burrow into the old leaf bases that protect the cycad stem and eventually cause the stem to fall apart. The larger, caramel-coloured *Phacecorynes sommeri* is less common than *P. variegatus* and feeds on the central part of the stem. Both beetles are usually only a problem when plants are stressed, for example after transplanting or when they are not properly watered (too wet or too dry).

Beetles are seldom a major problem on cycad leaves in South Africa. Snout beetles in the genus *Amorphocerus* occasionally burrow into the base of the leaves (petiole) thereby killing the leaf. Some chafer beetles feed on the leaf surface resulting in a transparent brown mark on the leaf and a small grey snout beetle that is common in many gardens sometimes attacks the young leaves.

To control *Antliarhinus* beetles in the seeds, spray the female cone with a contact insecticide (eg Malathion) at the time of pollination (March – June). Spraying at other times has no effect.

Beetles in the stem are very difficult to control. When a plant is transplanted or otherwise stressed, dust the stem with a fungicide and an insecticide (eg Karbadust). Beetles on the leaves seldom need to be controlled.

Aphids, mealy bugs and scale insects. The cotton aphid (*Aphis gossypii*) occasionally feeds on young leaves, and soft scales (Coccidae) and hard scales (Diaspididae) can build up on older leaves. Mealy bugs (Pseudococcidae) can be a nuisance when they develop in the crown. None of these pests are specific to cycads so if they occur on other plants in a garden they could easily become a problem on cycads.

Soapy water is often effective against aphids and mealy bugs. Oleum will control scale insects, and Malathion can be used against aphids and mealy bugs.

Diseases

Cycad leaves may occasionally go brown as a result of pathogens but problems with disease and leaf discolouration are often the result of poor growing conditions or other forms of stress. Stem damage through bruising or overwatering will result in stem rot which is most evident when the leaves go brown and begin to sag. If overwatering is the cause, plants can sometimes be saved by drying them out, treating with flowers of sulphur and replanting in well drained soil.

Nutrient deficiencies also cause leaf discolouration. Signs to look out for are the following.

- Leaves with yellow spots or irregular yellow splotches indicate a magnesium deficiency.
- New leaves that are yellow when they emerge suggest a shortage of zinc.

Above: Die-back of young leaves caused by heavy infestation of gall-midge

Opposite above: Damage to a cycad leaf caused by low level infestation of gall-midge

Right: Caterpillar of the Dimorphic Tiger (*Callioratis abraxas*)

- Mature leaves become yellow before they die, but premature yellowing is a sign that the plant lacks nitrogen which is essential for good growth.
- Brown and shrivelled new leaves may be due to a deficiency of manganese but look for signs of insect damage (see gall midge, above).

A shortage of nitrogen and some micronutrients can be corrected by applying appropriate fertilizers to the soil. However, some micronutrients need to be mixed into the soil and it may be necessary to repot or replant the cycad in a different soil mix.

A mixed planting of cycads and palms, Nong Nooch Tropical Garden, Thailand

LANDSCAPING WITH CYCADS

The range of growth forms, colours and growing requirements in South African cycads means that they can be used in a variety of landscapes. The foliage and stems (of the arborescent species) are the cycad's most endearing features and these need to be shown off to good effect. The cones are also quite striking, but they appear only at odd intervals and should be considered a bonus.

When landscaping with cycads, it is important to examine your garden and ask the following questions:

- What shape and colour of plant do I want?
- How big will the plant be when it is fully grown or has a full crown of leaves?
- What kind of cycad would be most suitable (ie a single stem, one with suckers, an aerial stem or an underground stem)?
- Must the plant be frost-hardy?
- Will the cycad be in the sun or shade?

South African cycads occur naturally in a variety of habitats such as the edge of the arid Karoo (see table page 19), in coastal forests, on the slopes of the Drakensberg mountains and in the subtropical lowlands of Mpumalanga. Despite the obvious differences in their natural habitats, most cycads do remarkably well in cultivation, and can

Encephalartos longifolius in a rugged Eastern Cape landscape

Opposite: A mixed planting of *Encephalartos* species and succulents

tolerate conditions that are very different from those they experience in nature.

Species from arid areas, such as *Encephalartos horridus* and *E. trispinosus*, grow very well at Kirstenbosch which has a much wetter climate, and at Fairchild Tropical Garden in Miami under warm conditions. Similarly, species from high-altitude grasslands, such as *E. friderici-guilielmi* and *E. ghellinckii*, which experience cold winters and even snow, thrive in the subtropical conditions of the Lowveld National Botanical Garden at Nelspruit. Another example of the adaptability of species is *E. latifrons* which grows well at Kirstenbosch yet does not thrive under subtropical conditions.

It is, nevertheless, important to take some precautions in areas with extreme conditions. Where severe frosts are common, it may be necessary to cover plants that are isolated and exposed, especially pot plants. In areas with high rainfall, cycads should be planted only in well drained soils, and in very hot areas most cycads do better if they are shaded during the afternoon. The table showing optimal growth conditions (pages 18-19) will help you to identify likely problems in any specific geographical area but a cycad's leaves are also a good indication of the sort of conditions it can tolerate. Cycads with narrow leaflets (<1cm) come from open areas with hot summers and cold winters. They like sun and can tolerate frost. Plants with broad dark green leaves come mostly from subtropical areas. They will grow in sun or semi-shade, and can be sensitive to frost. Cycads with broad blue grey leaves come from hot and often dry areas. They like to be in the sun, can tolerate some frost and need less watering.

One of the advantages of using cycads in landscaping is that they have a regular growth form and it is possible to plan an area knowing what the cycad will look like as it matures. Remember, though, to choose the correct cycad for a particular area. This is particularly important when starting out with a young plant. A cycad with a large crown and the potential to develop a tall trunk can make an attractive feature plant in an open area especially when planted in groups. However, one of the smaller species with an underground stem may be completely lost in this context. The converse would be true in a confined area. Refer to the table of features to identify plants with suitable growth forms.

In general, cycads are well suited to raised areas within a garden. The extra height shows off their foliage and provides good drainage which is essential for them to flourish. They also look good when planted among rocks and those with large crowns can look stunning among quite large boulders. The species with blue-grey leaves are often associated with quartzitic sandstone in nature where the light colour of the rocks contrasts with the blue-grey colour of the leaves. Re-creating this effect in a garden is very rewarding and these plants can be used to good effect in rockeries.

Planting cycads with other plants can be used to create different effects. When grouped with agaves or other cycads, the linear features of the leaves are accentuated. This creates a strong visual impression. Softer effects can be created by grouping cycads with less well defined plants. Cycads with blue-grey leaves combine well with succulents in garden plantings and make attractive feature plants in rockeries.

Encephalartos lanatus in among boulders – note the attractive skirt of old leaves

CYCADS AND THE LAW

Do you need a permit?

The reason for having a permit system, and indeed for any conservation measure, is to protect cycads in their natural habitat. An important part of conservation is to encourage people to grow cycads from seeds (of garden origin) or from nursery-grown seedlings, and there are relatively few restrictions to these activities. The permit process becomes more complicated for trade in larger cycads (which could have been taken from the wild) and for international trade in cycads. It is impossible to provide a comprehensive guide to international agreements and permits in all nine provinces in South Africa in a book of this size. The permit system in the provinces is also currently under review to make it easier to own and trade in nursery-grown plants while using more sophisticated technology (various forms of markers) to identify wild-collected plants. As a result, a detailed description of current laws will soon be out of date. The essential steps you need to take to obtain a cycad legally are outlined below.

- For exotic cycads bought in South Africa: no permit is required.
- For seedlings of indigenous cycads bought from a nursery: Keep the receipt. In some provinces you may have to obtain a permit at a later stage as the

33

Confiscated booty – cycads that were illegally removed from the wild wait to be relocated to a safe site

plant grows, and the receipt is your proof of purchase.
- For larger indigenous cycads: Registered cycad nurseries or traders will usually obtain a permit for you. If you are in any doubt, contact the flora permit section in your provincial nature conservation department.
- For export or import of cycads/seeds: South Africa has ratified the Convention on International Trade in Endangered Species of Fauna and Flora (CITES) and the country is therefore obliged to regulate the export and import of endangered species. All South African cycads are listed in the category of highest threat, ie Appendix I. This means that to export South African cycads (or any part of them), you first need a CITES import permit from the receiving country and an export permit from South Africa. At present, only some of the provinces issue CITES permits so contact your provincial nature conservation department for details.
- To import cycads from other countries: for Appendix I species you need an import permit from South Africa as well as an export permit from the sending country. For Appendix II species, you require an export permit from the sending country.

How do you know that your cycad is cultivated and not collected from the wild?
This is not usually a problem for the average gardener or enthusiast who obtains plants from a reputable nursery. These plants are mostly less than 10 years

old and there is no doubt that they have been cultivated from seed. It is really only people who want to collect rare cycads, or who want large plants, that run the risk of dealing with wild-collected plants. It is important to realise that some plants were wild-collected many decades ago and now have valid permits. These plants can be bought and sold if the right permits are obtained. It is, however, now illegal to remove cycads from the wild or to trade in cycads that have been illegally removed from the wild. Nature conservation agencies are marking plants in the wild with microchips and other forms of identification so that they can trace plants that have been removed from nature. If there is any doubt about the origins of a plant, contact your provincial nature conservation department.

FURTHER READING

Donaldson, J.S. 1995. The Winterberg cycad. *Veld and Flora*, **81**(2): 36-39.

Encephalartos, Journal of the Cycad Society of South Africa. Membership: P O Box 1790, Groenkloof 0027, South Africa

Giddy, C. 1974. *Cycads of South Africa*, Struik, Cape Town.

Goode, D. 1989. Cycads of Africa, Struik Winchester, Cape Town.

Jäger, A.K. and Van Staden, J. 1997. Cultured cycads. *Veld and Flora*, **83**(4): 113.

Osborne, R. 1993. The cycad collection, Durban Botanic Gardens

Tang, W. 1995. *Handbook of cycad cultivation and landscaping*. W. Tang, Florida

Van Jaarsveld, E. and Welsh, R. 1995. In search of *Encephalartos woodii*. Veld and Flora, **81**(2): 40-43.

The Cycad Conservation Project

The Cycad Conservation Project was launched in 1992 by the National Botanical institute with the aims of promoting cycad conservation through the propagation and cultivation of threatened species. A research component will also determine priorities for conservation, the chances of survival for different cycad populations and the best strategies to ensure long-term survival of wild cycad populations. As research necessitates extensive visits to natural populations, a vehicle dedicated to this purpose was provided by the Mazda Wildlife Fund.

If you would like to help sponsor the programme, which has developed close links with other organisations involved in cycad conservation including the Botanical Society and WWF-South Africa, or would like more information, contact Dr John Donaldson, the Project Leader: tel. (021) 762 1166 or fax (021) 797 6903.

The Mazda Wildlife Fund vehicle at an *Encephalartos trispinosus* **population**

INDEX

Page numbers in italics denote illustrations.

Amorphocerus 27
Antliarhinus peglerae 25, 26-27
Antliarhinus zamiae 25, 26-27
aphids 25, 27
    Aphis gossypii 27
Callioratis abraxas 25-26, 27
    beetles 26-27
    caterpillars 25-27
CITES 34
compost 17, 20, 21
cones (male and female) 6, 8-14
conservation 6, 33-35
container plants 23-24
    growing medium 23-24
    position 24
    watering 24
cultivation 17-22
    feeding 21
    planting 17
    soil 17
    transplanting 21
    watering 20
Cycad Conservation Project 35
cycad weevil 25
cycads
    availability in nurseries 18-19
    centres of distribution 5
    distribution in SA 18-19, 29
    families 5-6
dimorphic tiger moth 25-26
diseases 25, 27-28
division 15-16
Encephalartos 5
    aemulans 4, 18
    altensteinii 17, 18, 20, 23, 24
    aplanatus 18
    arenarius 18, 24
    brevifoliolatus 18
    caffer 5, 9,18
    cerinus 18
    cupidus 18
    cycadifolius 18, 21

    dolomiticus 18
    dyerianus 18
    eugene-maraisii 18
    ferox 18, 22, 24
    friderici-guilielmi 18
    ghellinckii 5, 18, 21, 24
    heenanii 18
    hirsutus 18
    horridus 7, 14, 18, 24
    humilis 18, 22
    inopinus 18, 24
    laevifolius 19
    lanatus 19, 21, 32
    latifrons 19, 24
    lebomboensis (cover) 10, 19, 24
    lehmannii 19, 24
    longifolius 5, 8, 19, 20, 24, 30
    middelburgensis 12, 19
    msinganus 19
    natalensis 19, 24
    ngoyanus 19
    nubimontanus 19
    paucidentatus 19, 21
    princeps 19, 24
    senticosus 7, 19
    transvenosus 19, 24
    trispinosus 19, 24, 35
    umbeluziensis 19, 24
    villosus 8, 15, 17, 19
    woodii 2, 19
exporting cycads 34
fertilizers 16, 21, 24
frost 31
fungicides 12, 13, 15, 27
gall midge 26, 27, 28
geometrid moths 25
growing conditions 18-19
growing medium 23-24
importing cycads 34
indoor plants 23
insecticides 26-27
insects see pests
Kirstenbosch cycad collection (cover)
landscaping 29-31
law see permit system

leopard magpie moth 25
loopers 25
manure 17, 20
Masson, Francis 5
mealy bug 25, 27
midges 26, 27, 28
mulching 17, 20, 21
natural habitats 18-19, 29, 33-35
Nong Nooch Tropical Garden, Thailand 28
nutrient deficiencies 27
offsets see division
permit system 33-35
pests 25-27, 28
Phacecorynes sommeri 26
    variegatus 26
planting 17-20
pollen collection 9
pollination 9-12, 11
pot plants see container plants
propagation
    medium 13, 15
    pollination 9-12
    seed 9-15
    vegetative 15-16
scale insects 27
seed
    germination 12-15
    harvesting 12
    propagation 9-15
    sowing 12-15
    storage 12
snout beetle see weevils
soil 17, 23-24, 31
sowing mixes 15
Stangeria 5
    eriopus 6, 19, 23
suckers 15-16, 18-19
Thunberg, Carl Peter 5
transplanting 21
watering 18-20, 24, 31
weevils 25, 26-27
wild cycads 33-35
    see also cycads: distribution in SA
Zerenopsis leopardina 25